To Rose Wheeler
whom I —
admired as a brilliant
psychologist and social
worker.
I hope you enjoy the
poems in this collection,
including mine. The best to you!
Esther Kossoff

SOUND and LIGHT

POETS CLUB OF CHICAGO
1937-1987
50th Anniversary Anthology

LAKE SHORE PUBLISHING
373 Ramsay Road
Deerfield, Illinois
60015

Lake Shore Publishing

Editor-in-Chief	Carol Spelius
Managing Editor	Wayne Spelius
Comptroller	William Spelius

(Poets Club of Chicago)

Assistant Editor	Agnes Wathall Tatera
Assistant Editor	Gertrude Rubin
History	Ann Nolan
History	Gladys Cook

Cover Artist	Ronald Crawford

Library of Congress
Catalog Card # 86-083171
ISBN # 0-941363-00-7

Table of Contents

POETS IN MEMORIAM

Blanche René, 1986

Rita Duskin, 1986

Louise Hajek, 1985

Wilma Burton, 1982

HISTORY
OF
POETS CLUB OF CHICAGO

Fifty years ago, three Chicago high school teachers made a formal club out of their gathering of a small circle of writing friends. Hence was born the present organization of Poets Club of Chicago.

The club has met in numerous places over the years: the McCormick YWCA, various rooms in the Fine Arts Building, a courtesy meeting room at the Uptown Savings and Loan on Michigan Avenue, and presently, at the Newberry Library.

Poets Club of Chicago has supported poetry contests given by Midwest Writers Conferences and donated prizes. The club also sponsors an annual international Shakespearean sonnet contest, and has given birth, full-blown, to the group of over 200 members known as Poets & Patrons Inc., who run up to 22 annual poetry contests yearly, open to Chicagoland poets, with cash awards.

Poets Club of Chicago has remained, itself, a small body with around 30 members. Prospective members are usually invited to join, and must present work to the committee, and the work and poet must be voted on by the membership.

The club meets once a month, each member bringing at least 15 copies of one of his poems. After a short business meeting, poems are passed around and each poem is read aloud several times and criticized. Intelligent and creative criticism is the rule, and the strength of the group is based on gentle helpfulness.

The current Poets Club of Chicago president is Agnes Wathall Tatera. Former presidents include Isabelle Gillespie Young, Lillian Foster, Helen T. Brown, and Don Cornwell.

MARGINAL IMAGES

never cross the tv screen
or jump in your lap
  and lick your face
never grab your hand
and say hello.

Marginal images
live in a place
  of old songs
nursery rhymes
in the blurred outline
  of someone gone
the grey streak
  of a lost cat.

Marginal images
catch in your throat
stop time short
  as faded photographs
crawl into bed
  after you're asleep
follow you like shadows
  in the corner of your eye.

...SPOON RIVER QUARTERLY, 1985.

THE LAST ACT

The marquee blinks a bloodshot eye
  at tenants passing in and out --
  old theatre people looking through
  each other at imagined rainbows.

The Flaherty sisters, in grey sweaters,
  watch cabbage roses wilt off walls
  in the old hotel.
Dust, in thin sun rays, spotlights
  cartons, bottles on a window ledge.
Geraniums, dying, take final bows.
Costumes with spangles, feathers and fur
  moulder like dead animals. Pictures
  of smiling girls shimmy from frames.

Jack Henderson and his "sweetie" empty
  their jar. Pennies fall like petals.
She looks at his greying tux and longs
  to be sawed in half once more.
Smoke halos "the funniest guys this side
  of Pittsburgh". One shuffles and deals
  while the other frowns at the cards
  as if they were naughty children.

At reception, Ben reads the racing form,
  yawns. All is quiet except for pipes
  hissing and spitting like cats.

Gargoyles grin from cornice and lintel
  festooned with scrolls and leaves.
Bay windows eye the street
  for yesterday's stars.
Paint peels, spiders spin, decay lives
  at the St. Elmore.

...*OVERTURES, 1986.*

STREET DANCER

He is a tuxedoed geometry
of waving arms, tapping feet --
a marionette on invisible strings
dancing in the theatre of the El.

Mirrored glasses kaleidoscope
flower vendors, candy men, passersby.
A woman drops coins into a hat
as if she were genuflecting.

His spectator shoes beat a rhythm
to swing hollowing from a silver box.
Black, white, black, white,
a blurring, whirling, turning
puppet of light and dark.

Peripheral people hurry by,
are swallowed up by trains.

The King of Swing stops.
Flashing red suspenders,
the street dancer bows,
turns, greys into the crowd,
leaving light behind.

...SALOME, by Ommation Press

PLAYING SOLDIER

Abdul says I shoot as well at twelve
    as any of the great Jihad.
In my gray fatigues and soldier boots
    I am important
as any rooster in the town square.

But this priest I guard
    makes me feel bare as when
I was a baby, staring at me
    with a father's eyes.

I want to pluck a bullet from my belt,
    show him what I can do with it,
make him stop talking to the Jesus-man.
    They say he is the enemy of Allah.

If I do this thing --
    close the priest's eyes forever --
will Abdul let me clean the guns,
    maybe give me chocolate?

...First prize, contest, JEAN'S JOURNAL
Published in JJ, March 1987

Constance M. Vogel received her B.S. in English from Marquette University, M. Ed. in Language Arts from Northeastern Illinois University. Her poetry has been published in ORPHIC LUTE, SPOON RIVER QUARTERLY, OVERTURES, JEAN'S JOURNAL and other poetry magazines, as well as NATIONAL POETRY ANTHOLOGY, 1982-83.

Glenna Holloway

LAST UNCLE

In my dandelion days
I had five tall men
who swooped me to their shoulders,
swung me, made me squeal deliciously.

In my frantic floribunda days
they began leaving
and I let them, without ever learning
of the last year of the Cavalry,
without feeling the bees' warmth
against the glass lab cases,
meeting the queens, tasting
the special honey, without watching
the wet hands at the potter's wheel
but always reaching out my own
for prize-winning pots.

Once in my pastel hollyhock days
it served me to serve with the cognac
what I had heard without listening --
how one had flown
the early U.S. Mail, had twice met
Lindy, had crashed in a swamp but
saved his cargo. A dry-box tidbit
with only a faint aroma of rich
ragout. Because I was always
singing in the choir, collecting
carnival glass, photographing
children in the garden.

(cont.)

And now in my tearose days
there is one. My faded luggage
is crammed with empty pages
as this jet closes an east to west arch.
And I wonder if he'll feel like talking.

...POETRY ANTHOLOGY, 1980
National Federation of State Poetry Societies

BUTTERFLIES AND OTHER NOBLE THOUGHTS

What about those that wobble wet and wingless
on the torn flap of the chrysalis?
Born too early or late, too much, not enough.
Some do soar via bird beak and maw;
most fall unknown in the ragweeds
to be savaged by ants before flying as dust
in the jaws of prominent winds.
Do their glistening granules return
to incubate again in more fertile capsules
or must they wander wasted
forever looking for their missing colors
and a womb?

...PRIZE POEMS, 1979,
Pennsylvania Poetry Society, Inc.

THE INTERLOPERS

Beneath inverted black jungle
of water hyacinth roots underweaving
	my hidden bayou,
my diver's lamp the only hold with
	my world,
I disturb a concert of stripes: hundreds
of inch-long fishes silver-slanting
	right or left
as my hand directs. A king-size mud cat
like Genghis Khan eyes me from
	the olive drab floor.
And overhead! My lost boat! Impounded
since last summer -- clamped listing
	in a wet/dry vise,
sun-half of bulbous green vases feigning
innocence with flowers; night-half
of fringe and garland chain,
propeller upholstered in velvet.
I rip away the slimy grip and feel
hairy stalactites creep closer,
more determined than topside kudzu.
The gasoline-fed screw might
thresh a yard before losing.
A new spring army of trees
wades out to make a stockade.
Roman-helmeted herons patrol
the spreading perimeter above
	with lances.
Here, the mighty Khan rules,
guarded by turtles. And I,
slave to light and lungs
must fight myself free.

...DIAMOND ANTHOLOGY,
Poetry Society of America

(While I'm Trying to Write a French Rondeau) YOU ASK IF I'VE FORGOTTEN HIM

I think of him when lightning splits
The pines astride this ridge then hits
The thicket wall still rolling fire.
Brush flares, rain hisses, makes a spire
Of smoke, a claim. It rarely quits
With that, but smolders in the pits
Of sticks and limbs till all my wits
Are stoked, so after I retire
I think of him.
In warming times each year this blitz
Re-scars my woods. I search for bits
Of green around charred wounds,
then liar-
Like, hail life as something to desire.
All winter when my likeness fits,
I think of him.

Glenna Preston Holloway has had prize poems in NFSPS every year since 1977, has published in GEORGIA REVIEW, WESTERN HUMANITIES REVIEW, POET LORE, PUDDING magazine and many others. She serves on poetry panels, gives readings. She has studied with James Dickey.

Gertrude Rubin

CONSPIRACY

Midnight, late August, I'm still awake
not in bed, but chilled on the porch,
waiting for a few friends to share
the night: for example, the prestigious

Dipper I earlier saw, north/northwest,
behind the yard's sloped garage.
Where did the moon go, after its face
was fully unmasked? Vanished like

Venus, the planet I need to steady
my course. From this perch, the sky
is a half-deserted room where, I swear,
something was whisked, dragged away,

seconds ago. A sea of tranquility but
contrived, shot through with holes --
among the missing I count Dog Star,
Hunter, Queen in Her Glittering Chair.

Late stars are more arrogant with
the sky to themselves. Like perfect
strangers, they stand together, cold-
eyed, distant, defying my losses.

A.M. SCENARIO

It begins with a jet plane's
roar. Houses line up
in tight-lipped silence.
Dreamers stir uneasily,
like flotsam swept ashore.
Thunder. Distant timpani.

From fog, trees emerge.
They stiffen if catastrophe --
fire, plunder, death, divorce --
hits our block. Or if the SUN,
flung from a newsboy's wagon,
misses someone's porch!

Churning trucks. Bravura whistles.
Garbagemen alight, and
lift the cans like props.
Next comes the early jogger,
careful not to look back
at the thing pursuing him.

So the stage is set:
alarms rung, "curtains up",
and Morning greets us,
blushing, on cue. Preceded
by a cacophonous chorus of
horns, lawn mowers and birdcalls.

LITTLE THEATRE ON HOWARD STREET

When sunset's backdrop falls
and night upstages day
kaleidoscopic signs
erupt like stars ablaze
their glare illuminates
the street of no return.

Undaunted by the past
new storefronts bravely rise
as late October rains
expunge the street of sin
and life transformed by art
opens at Wisdom Bridge.

In Brian Friel's play
a Healer of the faith
condemned to ply his trade
to Celtic countryfolk
is nightly crucified
by those incurables.

The sickly and the weak
(a charlatan's delight!)
defrauded by his touch
escape with ghostly steps
across the narrow heights
the depths of Howard Street.

And yet in crowded rooms
above the juke-box bars
men dressed in undershirts
unlock their steel guitars
while overlooking grief
they strum a salsa beat.

(cont.)

Their voices drift along
the corner fiddle shop
where instruments in view
are carved and rubbed by hand
raw wood once virginal
conceivable of song.

CELEBRITY

You, the alien angel,
awaken to thunderbolts
of alarm; warmed-over
maelstrom for breakfast.
A mirror, its vacant eye
of daylight, confirms
your wound. Stuffing
your tattered plumage
in a frogman's wet-suit,
you flop through streets
of vaguely swelling ocean.

Gertrude Rubin has given poetry readings at Northeastern Illinois University in Chicago, the Skokie Public Library, and at several senior centers and nursing homes in Chicago and suburbs. She has been a Dial-A-Poem-Chicago! reader; has had poetry published in small presses and in the 1985 ANTHOLOGY OF MAGAZINE VERSE & YEARBOOK OF AMERICAN POETRY.

Gladys Cook

THE SILENT HORN

Billy, the one I loved, loved music,
jazz music.
He worked in an office;
he lived for the sounds
that spilled out of his trumpet.
Billy soloed with big bands ...
on the record player, I mean.

My man never needed a note
          he didn't have;
he kept them there inside his horn
and he would lift each one up
          and hold it,
rounding it out and filling it in
before he let it loose to go its way.
Sometimes he let the music flow
clean and pretty;
sometimes he blew a handful
of sharps and flats
so furious and fired up
they warmed the whole room.
When he worked and couldn't play
          his horn
Billy listened to music he carried
          in his head.

His trumpet died with him
          thirty years ago.
Billy would have hated being old,
anyway.

...THE PEN WOMAN, 1984
...POETRY NORTHWEST, The Oregonian, 1981

MAGNOLIA TRILOGY

Part I: voice of SISTER

Brother, come!
Tune me in some good music
like they played when we were young.
I still hear that band
at the Mardi Gras Ball
thirty years ago.
There you go again saying it was fifty!
Did my mammy wheel me there
in a go-cart?
No! My husband took me
and him old as our daddy.
How I charmed the young men that night,
danced and danced while Charles
went sulking, jealous old man!
You dial too fast, you passed
a good station.
The announcer sounded like
the French Consul at the ball.
When we waltzed, he said
I weighed no more than eiderdown;
I was always weightless on my tiny feet.
Must you leave?
You never want to stay and talk.
I suppose you do have work to do
in the kitchen but dial me
another station first;
that music is not like New Orleans.

Part II: voice of BROTHER

She is my albatross,
    my addle-pated sister.
Penance I have done and penance more
shall do for in my heart
I've killed her many times.
Once she was small, now she
spreads herself over the chair seat,
dips her fingers in the sweet box
and clamors, "Brother, come!"
Foolishness is her fetish
and ever has been.
She flirted with the young men
until they were aflame
but she married old Charles
who died and left her to me.
Oh God, her conceit! Glasses conceal
her charm so she teeters
through dimness on spike heels
and topples to the floor.
Old bones don't heal. . .
I do the woman's work while she calls,
"Brother, get me some good music."
I am old. If I go first
who will take care of my poor sister?

Part III: HUSBAND

(voice from the grave)

"Loving Husband." Words chiseled
 in stone.
"He was a fool," is truer epitaph.
Like the Spaniard, I thirsted for youth
and she was young - so young!
Surely, she'd share her youngness
 with me.
I wooed her with gifts beyond my purse,
visioning enchantment on the road ahead.
The altar march was short --
a short cut to my grave.
An old man wearies wed to
a new-fledged lass;
he lies sleepless through the night
pondering ways to buy her
pretty clothes.
Later, when jealousy becomes a disease
that chews his entrails,
he is thankful for peace
even though the stone above him reads,
"Loving Husband."

...THE HUMAN VOICE, 1969

Gladys Cook graduated from Columbia College at age 60, in 1967. Has books published here, in Finland, and in Germany. Many of her poems have been published and awarded prizes. A former Chautauqua actress, she has done many highly successful poetry readings.

Catharine N. Fieleke

SCARECROW

On the far side of Hell
a scarecrow figure wanders
face streaked with soot
garments scorched and tattered.

Eyes blink and blink
shy of the enfolding light
feet move slowly
troubled by the resilience
  of grass.

Brook says to Scarecrow
  <u>splash your hands and face</u>
Breeze says
  <u>have no concern</u>
  <u>I will carry away your stench</u>
Clouds in the blue say
  <u>listen! serenity is real</u>
Sun says
  <u>brother, your bones are cold</u>
  <u>lie down, let me make them warm</u>

Slumped in the cradle of the
  earth he sleeps

Now will come the sheriff and
  his assistants:
  You there, up on your feet!
  No loitering permitted.
  Off with you, move!
(cont.)

Or -- a cowled priest will pass
saying his orisons
automatically making a sign
  of peace.

Or -- a child, silent, amazed
will circle by and suddenly
  take to his heels.

Or -- could it be that a farmer
  out for haying
balancing on his shoulder a scythe
  and a rake
bread in his pockets, jug in his
  burly hand --
could it be that he will retrace
  his steps
stand staring
sorting his thoughts in the
  urgent morning sun?

  Shall I rouse 'im?
  What if it rains today?
  If I fed 'im
  could 'e work the rake?

Oh, grant all scarecrows such
at least this such
grant their poor stick bones
the dreaming, fitful boon
of meadow sleep...

SASSY CATBIRD MONOLOGUE

Miaow! Miaow!
I like who I am --
That's why every five minutes I
  announce my name:
Catbird! Miaow! Miaow!

When the powers that be finally
  realized
I have a splendid voice
They set me to study under
  Mocking Bird.
What a tyrant!
Work work work work work!
Miaow rests my throat from
  all those flourishes!

If you want to know my real
  philosophy
I say -- be yourself -- let it
  all hang out!
I'm Catbird. Do the best I can --
Most folks seem to think that's
  passable -- in fact --
  pretty good!

Ooops! Gotta go now -- time for
  more tr-r-r-rilling!
Miaow! Miaow!

Two collections of Catharine N. Fieleke's poems have been published, and two are pending.

DIALOGUE ENDING IN A QUESTION

Let me get this straight;
You're saying there's a law
of opposites
a cosmic tensing
against negatives.
Well look:
male female
dark light
cold hot.
What's new
what's sun-up about that?

Matter anti-matter.
Are we
proof
of angels?

...SPOON RIVER ANTHOLOGY

Catharine Fieleke's poetry has been published in LINCOLN LOG, VOICES INTERNATIONAL, WELLSPRING, ENCORE, EARTHWISE CALENDAR, THE OREGONIAN, THE VIRGINIA QUARTERLY REVIEW, CHRISTIAN SCIENCE MONITOR,THE STANDARD, MIDWEST POETRY REVIEW, and is included in several anthologies. She was for four years Poetry Editor of THE PEN WOMAN, authored 22 scripts for classroom enrichment, the contract for distribution running for thirteen years.

Donna Dickey Guyer

CHILD WITH A BUTTERFLY

Oh little fingers,
let the bright wings go.
They have so small
a time in which to fly,
and here is something
that you soon will know --
if butterflies are held
too long, they die.
Learn now to catch
the moment's flame, young eyes,
and satisfied with looking, turn away.
There is no end to wanting butterflies,
no boundary for trails
too far from play.
Be wary in pursuing stars, young feet,
for pathways wind
and sometimes disappear,
and radiance is often incomplete
when captured. Be content
to linger near.
Truth hidden in a paradox is vast --
releasing beauty, you will hold it fast.

...THE LYRIC
...Winner of Lyric Reynolds Award

Donna Dickey Guyer has a chapbook of poems on Chicago to be published in 1987, by Lake Shore Publishing.

WHITE PELICANS

To grace one moment with a feathering calm,
fixing a shard of time against all space,
an arrow of white pelicans arched by
soundless in flight, immeasurable in balm.
And I was lifted from the commonplace
to make my intervention in their sky.
Slowly they soared until the hungering eye
feasting upon the winged dots, lost trace.
And then a whiteness floated on the mind,
moment of freedom from the day's embrace,
as though the world released me from its palm
to follow where white pelicans inclined,
able to keep the spirit intertwined
with body, in the lifting of a psalm.

...THE LYRIC
...Winner of Lyric Memorial Award 1985

Donna Dickey Guyer has won Lyric Memorial and Nathan Hale Dole prizes, and was Poetry Editor of THE PEN WOMAN, official magazine of the National League of American Pen Women.

YOU CAN SEE THE OCEAN FROM THIS WINDOW

(Nursing Home)

Mind is pain-scarlet
when old bones sleep.
Colors are all you can hold
onto any more.
Sound is yellow bursting
as the sea flows in,
and touch is brown,
fingers kissing a lost world.
Blue's death --
you can't smell blue.
Tasting is ebony and sight's white,
but the waking mind is joy-green
when old bones know the ocean is there,
outside the window,
waiting.

...PORTLAND OREGONIAN

Donna Dickey Guyer has been included in several anthologies, and has an essay on writing dramatic monologues in a current textbook, "The Study and Writing of Poetry: American Women Poets Discuss their Craft."

TROPIC DRIVE

Twist of trees
and the wind's whip,
the sea's white snarl,
and the lonely widows
steering their cadillacs
to nowhere
among the condominiums.

Rinse of rain
and the shuffled shells,
the sky's hot yawn,
and the bent old men
walking their poodles
to paradise
along the scenic route.

...HYACINTHS AND BISCUITS

Born and raised in the Windy City, Donna Dickey Guyer has a B.A. degree from the U. of Chicago. Started writing at the age of twelve. Has had poems in the CHICAGO TRIBUNE, WRITER'S DIGEST, READER'S DIGEST, NEW YORK TIMES, DENVER POST, WALL STREET JOURNAL, THE LYRIC, McCALL'S, LADIES HOME JOURNAL,CHRISTIAN SCIENCE MONITOR, etc.

Gertrude Robb

VILLANELLE TO HELEN

I swear by candle, book and bell,
I swear I do, and cross my heart,
I cannot write a villanelle.

I can not weave by witch's spell
delineated poet art,
I swear by candle, book and bell.

I prayed I'd never have to tell
I cannot even make a start.
I cannot write a villanelle.

My facile pen cannot excel
by arbitrary rule and chart.
I swear by candle, book and bell.

I swear by all the hounds of hell,
by Jacob's beard, and Cupid's dart,
I cannot write a villanelle.

Enough's enough. And hear me well.
Despite the wisdom you impart
I swear by candle, book and bell
I cannot write a villanelle.

CALLIOPE

When that which I have been
shall cease to be,
or, snugly shrouded in a greyed cocoon
of circling self,
seek temporary flight
from this mod galloping calliope,
let there be only stillness.

Let no tune or sybaritic sound
disturb my night of breathless sleep;
no woeful acolyte, with surpliced song,
profane
nor yet buffoon, with tattered bells
and flourishings,
protest the quiet hour.

Morning comes too soon when one
is weary.

If, at reveille,
I soar angel-winged
amidst the blessed,
or loll in nothingness, unwound,
unstressed,
it will not matter overmuch to me.

THE SILVER SADDLE

Death rode forth in his silver saddle.
Breathless, the angels saw him
depart,
for the Saddler had cushioned
the silver saddle
to cradle the baby I held to my heart.

Death came prancing. A jingling
silence
fell like a pall on the golden sun,
Though the silver bells of the jingling
silence
were tuned to the need of my little one.

So limpid the lure of her latent
listening,
so fragile the heartbeats for life's
demands,
I muted my love to her latent listening.
Science stood helpless,
with waiting hands.

Death rode high in the grey day's
dawning
rearing and plunging
with blood-stained spur.
Mute and grey as the grey day's dawning
the angels of heaven looked down on her.

Death rode high in his silver saddle.
I prayed to The Saddler with lips
of stone.
Prayed to the God of the Silver Saddle.
Death rode back, alone.

SHOUT OF FLAME

You need not walk with me.
This narrow lane were pain to you
who still would know the eagle view
of crag and hill.

You hear the shout of flame
                    on mountain top,
and you must go.

I am content to know the gladness here,
the gentian's blue,
and flitting hue of grasshopper
on grass.

I cannot share your booted climb
to taunt of twisted pine and glacier burn.
I cannot follow you,
but I can wait
for your return.

That is,
if you are not too long away.

The gentians are a lovely blue today.

Gertrude Robb, a former school teacher who has been involved with Little Theater (especially for children) has received many awards for her poetry in national, international, and local contests. She has lived in the Chicago area for many years.

John Dickson

LYING ON THE GROUND

Studying clouds in their shifting about
or the way maple trees take root
        in the sky.
The honkings of duck trains hang
        in the air
and the cries of squad cars in heat.

But under the grass lies another world
-- cables roaring their buried secrets
over and over and all at once
of "Meet me down by the boat house
        tonight"
of "Ben, you hardly ever call"
-- man voice of chest hair
        and day-old beard,
smooth-skinned woman voice fading away.

And there's the turbulent rush of pipes
moving the lake from there to here
and the giant drains sucking
        and tumbling,
sloshing the endless sewage away.
Like resting my head on an animal's
        belly
listening to trickling waves of sound --
worms plastering their tunnels
        with slime
as they drill matrix paths
        around the world,
tree roots and rose roots feeling
        for life,

(cont.)

iris and jonquil bulbs swelling
            and pressing
deep in their darkness of slow birth.

I hear decomposing cats and dogs
once loved and living
            and purring and barking
in the surrrounding houses.
Or the sound of moles, of gophers
            and subterranean grubs.
And buried weapons -- ticking bombs,
rusting revolvers and bladeless knives
-- bullets lodged in collapsed skulls,
hair and nails growing and growing
as rivers of pain drain into the soil
down, down to the underground ocean,
souls locked up in caverns of clay,
weeping like sacred statues.

A giant magnet is the earth,
drawing down everything that lives --
leaves and trees and ships and birds.
A giant recycling machine,
giver of life and life's destroyer.
But I've had enough of lying here.
I should leave or I'll blend with
cannons and cars and pennies
and rings and watches and wine,
should go to the crowded downtown
streets of pushing and shoving
and laughing and swearing
where I'm supposed to be.

THE UNIVERSITY CLUB

Of course he must have been a member
though not the sort you'd care to know.
Obviously just some Felix Nobody --
large baldheaded man in wrinkled suit,
always near the window in the dining room
reading some book or other as he ate,
gazing down at traffic in the street
or listening to foghorns groaning
in the harbor.

Every night he ordered octopus --
the only one, what's more, who
ever did --
Octopus Flambeau with creole sauce
and wine,
Octopus Del Monico or Octopus Kiev.
Or casserole, croquettes, or chowder.
Voracious appetite for octopus.

Twice after they passed the rule
allowing women
he brought the same disreputable floozy
who drank too much chablis, of course,
and prattled on in such a senseless way
with catfish gumbo oozing from her lips.
Her fingers twined like ivy around
the glass,
her legs wound around his
beneath the table and she sang
some strangely silly gurgling song
much to the embarrassment of everyone.
(cont.)

The head waiter was told to speak to him
if he should ever bring her there again.
But for months he ate alone --
Octopus Almondine...
Octopus a la King...
occasionally swimming up from
the caverns of his book to look down
at the traffic or listen to the harbor,
often with tears welling in his eyes.

One evening she returned,
so common with her flimsy dress
            and dangling beads.
They laughed and drank their wine
and sang their song as she brazenly
wrapped his legs with hers.
But when the waiter finally approached
            their table,
bristling brows and iron jaw,
she made a little purple-inky cloud
and poof! they suddenly disappeared.

John Dickson made Honor Roll ('61 & '68) in Martha Foley's "BEST AMERICAN SHORT STORIES." Won two weeks in Italy for poem entered in English Div. of POESIA GRADARA, sponsored by the Italian Gov. His first book, VICTORIA HOTEL, received Friends of Lit. Poetry Award, 1980. Conducts Lighthouse Poetry Series at Northwestern. Published in POETRY, TRI-QUARTERLY, HARPER'S, etc. His most recent collection is WAVING AT TRAINS published in 1986.

Mary R. Torode

ON THE THIRD NIGHT OF THE FLOOD

Claws clung to our china berry trees
all night.
Winds rode ploughshares through the roof
of our salon.
Birds' wings howled the windows
of our kitchen.
In the pond of the bedroom,
snakes wore Grandmere's boudoir caps,
and green pears floated over
the steel bed.

Under a lace petticoat moon,
over the roof of our chicken house,
on our bellies,
we gripped the sides of that roof,
and each other, with fingers and feet
all night.

At a calm with no wind nor rain,
dawn moved in gingerly;
and we climbed down from the roof,
down to the chicken roost,
lifted the setting hens
from their nests
and lowered our faces into the warmth
in their feathers.
If the lull lasted long enough,
we would swim to the bedroom
and chase the snakes again
and grab another meal
of green pears.

THE ATCHAFALYAN

Fishing became a simple art for the
Atchafalyans. To freshen saunas, she
brought wood violets or swamp lilies, or
lawn azaleas. She dusted bisque
tureens, peeled purple figs.
Domesticity promotes curvature of shoe
heels. Curved heels define boundaries
of slovenly poverty. She lined up her
curved, discarded pumps in her attic, on
a rafter to taunt great floods of the
future to reach, to touch. At birth she
had been bonded to flood water warmth.

The Great Flood of Epiphany crested the
rafter; on its grand current her shoes
floated out her window. Out the window,
in her pirogue, she floated. Noah had
felt foam-crested joys of drifting,
drifting along with his shoes; watched
the water levee them. 10 1/2 pairs in a
straight row. All this not dictated by
the Frugal God to the Bible Writers!

For her the Good God chose to add a
swamp man, a tiny man with the panache
of his yellow lilied cap. He patted her
half pair of slippers, reunited it with
the 10 1/2 pairs in the row on the
levee. "Grand women tantalize eyes
accustomed to worshiping lustrous
silhouettes of licentious reptilis --"
Circumstance wise he must have saved his
discretion on top branches of a thorny
acacia.

She purposely mimicked the chastened
voice of Elizabeth of The Young and The
Restless: "Go get some wild onions,
far, far down the levee! While I
eviscerate this dozen fish." He formed
the prayer tower with his hands; "I'd
rather miracle, for us, loaves of
bread."

Note: The Atchafalaya River flows through South Central Louisiana, the Cajun Country. Marie Rose Theriot first perceived swamp light, near Bayou Chene, Louisiana, nine days before the pen-ultimate comet. With the energy of comets and of her people, the Cajuns, she layers her poems. With the vigilance of her husband, Clarence Torode, grinding gears to the ten thousandths of an inch, she hones her images.

Credits Mary R. Torode is most proud of are: NIT AND WIT; ANOTHER CHICAGO MAGAZINE; SOUTH, University of Florida; TWO TONE, University of Rhodesia; FIRE FLOWER, White Horse, Canada; OCARINA, Madras, India; BOUQUET OF POEMS, Rome, Italy; and NRG, a semi-annual publisher of abstracted abstractions.

SOUND OF GUINEA HENS

In geography
guinea hens kerchank.
In Mandaville
black and white polka-dot
organdy dresses
slide up and down
bouncing stomachs
of 'Tonette and 'Enriette.
Papa has enough front yard
for two hop scotches
and three magnolia trees.

Under the house
between the brick pillars
it smells of old alkali,
of mildew of million years,
"Simonette, if you bring me
a bushel of cape jasmines
and we hide in,
the geography will not find us,
but it will hear the hens
kerchanking.

Mary R. Torode lectured on her poetry and Cajun culture at Triton College, and with Agnes Tatera conducted a poetry workshop for the college. She's read her poems at Columbia College and various open mikes, and she's taught senior and handicapped citizens under a Chicago Arts Grant.

Virginia Real Nicholas

SOUTH OF THE BORDER

Come dance with me.
My eyes flash
dark under my sombrero,
my white smile zings
like a light whip-lash.

Come dance with me.
My feet dart
swiftly as blackbirds,
my broad red sash
makes happy breeze
as we sway and part.

Come whirl and twirl
to the tambourines
that quiver and laugh
like little spanked girls;
to the clackety-click of castanets,
in the glide and the spin
your heart forgets
to cry, to cry with the violin.

...THE LYRIC

LOST TO YOUR SIGHT

You can perceive the dawn, you know
          when dusk
drifts like a sooted snow upon
          the light.
But lost to your sight, loved faces,
          and the world
in form and color, live in your intense
remembering. And with a thinking touch
you reach into the void to shape
          known things
and mold them with percipient clay
          of words.

The color of the sea still comes
          in sound
to one who lives an empathy with green
that tinges the salt-tang
          of coastal air.
Warm touch of sun may daffodil your day,
or lilacs breathe their lavender to you.

Like waters, moon impelled,
          the tidal dark
draws back before your mind's
          inviolate light.

...First Award, Experimental Sonnet
Poets and Patrons
...THE LYRIC

THE WOODS AND SEASONS OF OUR LIVES

Always it is the woodland
rising as in sea mist,
deepening to day-dusk,
and this need for the quiet
                    and the sounds,
scuffling through years of leaves,
stretching the mind.

April songs the throats of thrushes.
The forever pipes of Pan call to us
and the flute of Eros.
Fey with listening
we scale tall turrets of trees
to ride the sailing craft of clouds.

A crow whittles his caw.
Trumpets of color explode
yet sadness undertones the talk of trees
and the bending of winds
stained with sumac
evokes our melancholy.

Always the woodlands --
star-bruised even in day-dusk,
we seek cool moon healing
but the moon is not risen.

A millenium of trees, century wise,
encompasses this desolation
when darkness tangles
and last bird sounds are hushed.

...First Award, NLAPW Chicago Branch

NICKY

If a poem had sinew, muscle and bone
one might make a likeness of you --
striding about
with your head tilted back,
eyelids at half-mast
over the curious agate blue
of your eyes,

    an image of you
    talking vehemently
    your arrogant nose seeming
    to ride astride your words.

But a poem is frail
to bear your strength
and the arch willfulness of you,
to brace your body, lean, hard,
moving toward love
like a gull breasting
breakers of the wind.

Virginia Real Nicholas has been published in THE LYRIC, POET LORE, THE CHRISTIAN SCIENCE MONITOR, NORTHWEST MAGAZINE of the Oregonian, and CHICAGO TRIBUNE. She has poems in SOUNDINGS and numerous other anthologies; won Best of the Best award in national competition of NLAPW, and a Best of the Best in Poets and Patrons contests. Virginia also won a chapbook award from Wings Press, and many national and regional awards.

Dorothy L. Wahl

WHEN CENSORS WERE CAESARS

Weavers of Gothic novels
emerged skilled chefs
concocting, serving dollops of
virginal fruit salad or
connubial cream and marshmallow
nestled on fresh foolscap;
despite fillips of filberts
crisp leaves wilted.

Inspiration ad libbed:
paprickle garnish,
provocative asterisks,
spurred, catered to, burnished
epicurean fantasies.

In fond remembrance
I humbly submit:
"Impatient, I yearn
for Caesar's return,
for an edible plate
that won't self-immolate."

URBAN SUNSET

A branding iron
singes through low clouds,
sweep-sears, ignites the bay,
detonates this rocky beach.
Maddered, peer stones
and I bleed furtive shadows;
boulders leech clumsy columns
surreptitiously.

Atop one,
squinting to eavesdrop
on ethnic wit of granite,
of quartz, I scoff, "The dead
  give thanks;

their life,
their speech is Majesty powered.
My fire flares within." Proudly,
arms outflung, I turn, see

my column, now
a spire, has engulfed my legs;
farther, a jutting rise
has guillotined my head;

shrunken,
it hangs ten yards beyond.

LILAC PSYCHOLOGY

There shall be thorny days, alas,
to stir a brief rapport with terms;
why agonize? They soon shall pass.

Our offspring, play-penned on the grass,
ignores his cookie, feasts on WORMS!
There shall be thorny days, alas.

His milk-teeth shed, we harvest sass
and skin more like a pachyderm's.
Why agonize? This too shall pass.

As a Mohawked teen, he's skipping class
to hang-glide in the highsotherms;
there shall be thorny days, alas.

Ensconced in brig by spiteful "brass",
...I'm innocent," our heir affirms;
why agonize? This too shall pass.

With cyanide -- for crimes en masse, --
he's spared slow death by smog
and germs.
There shall be thorny days, alas;
why agonize? They too shall pass.

DAISY

Moon-burst reflection: poised, calm
sun season injection: truth serum
no hidden guile
no overt wile
pride of earth, air
envy of sun, cumulus,
your eye, smile are one.

For your ethnic wind-dance
to click, whistle, pipe,
no panniered flounces
no layered petticoats,
you dip, bob with grace
slender leg bared
dainty foot clad
in giddy whispers.

Dorothy L. Wahl, of Arlington Heights, has had poetry published in TIME OF SINGING, a magazine of Christian poetry; SOUNDINGS, a poetry anthology; and ORBIS, an international literary quarterly. She directs a poetry workshop, writes lyrics for and composes sacred music.

Carol Spelius

INDIAN WOMAN AT HOSPITAL IN TAOS

I hear my heart beat above the strange
noise that echoes 'round me as if in
a canyon. All my bravery falls out
of my heart. Empty beats I hear.
The shiny floor glares back at me
and hurts my eyes.
Bad place, this House of Pain.

I know such fear only once before,...
back many winters when my braid
hung black not grey and I stood
closer to the ground than to the sky.
Then government men with guns took
my father and uncles away.

Now, not men with guns but my own
great-granddaughter, doll-faced
Kachina, took me to this place full
of cropped-hair devils, and to what
destruction I cannot tell. A strange
smell surrounds me. Not sage.
Not earth. A devil smell.

My mouth dries up with fright. I order
my arms and legs to lie still but
they tremble. I call to Kachina who
answers in the old tongue, mild as
mourning doves. "Trust me. These
medicine-men will make you well,
my foolish but venerated Old One."
Kachina is the foolish one.

(cont.)

Allows strange persons to tie down and
sting her great-grandmother who is
daughter of chiefs. Mother of chiefs.
I hold my face still as a pre-dawn slue
while my eyes shoot arrows at my
double-tongued great grand-daughter.

Kachina understands nothing.
Kachina teaches her son only English.
He does not sleep on the ground.
He does not wear mocassins.
He knows not how to be Indian.

In my chest burns a fire.
In my heart lives a sadness.
One more time I tell Kachina
to take me home. Not back
to foolish house where water
pours from walls and where earth
is not allowed to crawl inside.
Not that house. To the old house.
To the spirit house.
Where my eyes can find comfort
in colors of adobe.
Where my body can lie
in wool rugs on the earthen floor.
Where the sun can warm and heal me.
Where the spirits of
all my ancestors wait for me.

Kachina says no. Kachina knows not
mountains and desert long enough.
I howl inside like a trapped cougar
but no sound escapes my lips.
I think hard on the inside to pull
my spirit into the form of a hawk.

(cont.)

When time, I will soar to the sky
and head for home. I try not to cough.
The cough made Kachina bring me here.
They cannot see the fire. Only
the blood which always before
I wiped from my lips with my skirt.
Kachina scolded for that. She said
use the Kleenex, Old One. The Kleenex.
You want to make the great-
great-grandson sick?

I with arms and legs of a baby?
I with eyes that see only
in the distance and in the past?
I have no power to sicken even a spider.

Sounds dart around me like bats
in a cave. Strange hands touch me,
prod me. A masked man leans over me.
His voice does not frighten me now.
I stare into sun-bleached eyes
and think of sky.

He stings me, but like a mosquito.
I feel the hawk's beak pressing.
Dead snakes coil from the ceiling.
Many masks surround me. Very powerful
medicine. Good for great-great-grandson.
Not good for me.

The fire in my chest is only ashes now.
Desert and mountains loom beyond me.
I feel the hawk spread its wings.
I am ready for flight.

...NORTH DAKOTA REVIEW
...Mid West Poetry Festival, 1986

Wilma Burton

POSSESSION

Mine, the meadow! The cows that come
to graze,
dipping brown-white heads -- I claim.
The brook
that weaves its afghan-edge
to summer days,
the lark's cravat, tied in nesting nook,
are mine to view with
ownership's delight.
I watch him tilt on fences, arch
his song --
a shaft to splinter air
in measured flight.
These marvels I possess all summer long.
I see the sun spread amber marmalade
upon a hillock's crust of bread,
and share
with cow and bird
a winged-back rocker shade
on green shag-rug with fringe
of maidenhair.
A deed for land? O, who owns wheat
or rye?
First claim belongs to God,
then passerby.

SUNDAY WORLD

Suddenly is born -- a Sunday world!
Winter's work clothes lie
in sooted heaps
while willows, yellow ticker tapes
unfurled,
urge on a bull frog band that
marches, leaps
in uniforms of this year's green.
Earth's floor,
pomped and circumstanced
by last night's rain,
is now a promenade for barefoot lore.
Colt and lamb brave spindle-leg disdain.
When Sunday jonquils meadow, road,
should I
ignore the hours with ferret head,
in-bound,
deaf to larks on tilting fences? Deny
the fresh-bread smell of field,
of new plowed ground?

Creator with Created --
both are blessed
When Sunday's world
calliopes the breast.

HABIT WALL

I built a wall,
a habit wall
of happiness:
The darkling
things of world
shut out --
but it was lonely
inside, until I opened
the door a crack
and let another's grief
come in.

Whenever WILMA BURTON felt the spirit of inspiration -- day or night, at home or away -- she wrote poetry. Her work has appeared in CHRISTIANITY TODAY, CHRISTIAN LIVING, IDEALS, and other magazines. Her books of poetry are I NEED A MIRACLE TODAY, LORD, SIDEWALK PSALMS, and SOME FROM COUNTRY LANES. She died in 1982.

Joan Marella

RING NEBULA, M57

O eye
which appears to stare most coldly
but whose wide black pupil
floats in fiery rainbowed iris;

nebula, O nebulae,
your stellar winds rise
to blow out pieces of iris
as atoms vibrate
in absence of vitreous humor,
as dancing elements prepare your death.
Do you see,
staring at this blue planet?

Catch me looking out, looking
at you to discover the secret silent
space
created
at your final explosion.

RITES

Along the Ganges in chosen places,
the humans and beasts
wait one upon the other...
trumpeting and shouts crowd the ear;
bright beads of spray solemnize
the elephant bath,
the Hindu baptism.
Count the heads among the trunks,
each in turn raised
then dipped in mud-swollen holy water,
mud carried long after
on cracked skins of gray giants,
on ragged robes of slender souls.
Raman! In the morning brown river
wash these ceremonies together
to the unmuddied sea.

... ILLINOIS STATE DIVISION JOURNAL,1985
American Association of University Women

PICASSO'S LOVERS

Picasso's lovers overhead,
a simple pose in grays and blue,
give wry perspective to our bed.

My love what do you think they said
as artist brushed and likeness grew?
Those lovers hanging overhead

might tell where secret kisses led
and earthy wisdom shared with few
give wry perspective to our bed.

When they were hung, I lightly pled
if we enjoyed their knowing view,
Picasso's lovers overhead

could intimate how love is fed.
Is his caressing hand a clue
to wry perspective for our bed?

Would loosely buttoned blouse instead
of clinging silk make you pursue?
Picasso's lovers overhead
give wry perspective to our bed.

...ORPHIC LUTE, 1984

FAIRY TERN

A gossamer
called fairy wings
survives the salt
and whipping stings
of wave wash,
of wind whirl;
this nestling
hatched -- clings
stays, feather curl,
and sharply brings
the circle round
a retinal image -- turn
of light,
the fragile blur
of soundless white.

Joan Marella, most recently published in THE LYRIC and ORPHIC LUTE, has taught English and conducted poetry workshops for young people and adults. Her loves include family, writing, and music. She is a member of the Association of American University Women and president of the Children's Home and Aid Society of Illinois, Elmhurst Auxiliary.

Anne Nolan

AUGUST MADNESS

The day was a shimmer of tasselled corn
 Red-gold where the heat waves hung,
And the hours stretched out
 like a dozing cat
 With a field mouse taste on its tongue.

Back of the town square
 the tethered horses
 Shivered their flanks at the flies,
And the old clock marked the siesta time
 With alert, disapproving eyes.

Mother was washing...
 one tightly wound spring
In a world that appeared run down,
When suddenly, like an enormous pinch
 Or a swat on the rump of the town,

The circus train ground to a noisy halt
 And the ramps bit into the dust,
While the notes of calliope
 winged the air
 From the tracks where the houses rust.

The afternoon shattered to shouting boys
 Dog-chasing the news around.
Mother leaned to listen,
 almost as though
 She had hungered all year
 for the sound.

(cont.)

As her soap suds drifted
like lost balloons
And the plump sheets flapped
their surprise,
We followed that lilting, enticing song
To the field where the tent poles rise.

While the canvas strained to its
flagged-topped height,
Mother's budget went off
with the breeze.
She bought kewpies and treats,
center-ring seats
And a laugh-time of memories.

SONG TO MARY

The lullabies of birds lie sweet
Against the garden's breath
While one small soul goes tiptoeing
Across the land of Death;
And I am barren now, who knew
A brief, full mother part
With cradle songs at candle time
And curls against my heart.

And one small soul goes on alone
Who knew my hand before...
Dear Mother Mary, to your arms
I give the son I bore;
And as the twilight hours come
Peace...and my heart in prayer
Knowing my tired baby sleeps
In arms as gentle there.

NOT INTO THE ENDLESS DARK

Finally
he is gone

beyond the chant of clergy
and the futile tears,
the constant probe of needles
into tired old veins.

Somewhere
a little road,
daisy washed
and laughter freckled,
runs through the pasture
and the morning
to a secret place

and he,
boy-young again and
fish-pole ready,
knows.

LISTEN!

Over this wall around you,
word upon word,
doubt and distrust have found you.
Over this wall around you,
listen to love ... a sound you
have often heard
over this wall around you,
word upon word.

ESCAPIST

When all my lesser wishes
have been packaged
Like so much grocer's sugar,
pound by pound;
With social trifles stacked away
from exits
And day dreams tagged and binned,
wherever found,

I shall have little left of time
for wanting
Your arms about me in the hurry here.
If one can fill a life
with cluttered nothings
There will be work enough to down
the fear

That, someday, all this haste will end
in quiet.
Then I must turn to face
the road behind,
Too worn to fight the onrush
of my longing
Or still the frightened crying
of my mind.

Anne Nolan's prize-winning poems have been published in the CATHOLIC WORLD, MIRACULOUS MEDAL, EXTENSION, ST. ANTHONY MESSENGER, THE AVE MARIA and many others. She has won many awards and is a former president of Poets and Patrons.

Helen T. Brown

IMAGE OF A SNOWFLOWER

He was an old man and bemused
but he recalled her face -
her hair a burnished aureole
of silk-spun filaments of April fires
her eyes the lilac stealth
of young desires.
Was her name Helen
or did another archipelago
of sounds tremble with ardor
and with sudden stroke
convert its loveliness to briefer note?
Such beauty ravished Troy
a thousand stallions raced Olympic seas
to read her lips
then awed to silence by the hand of fame
paused in true homage to imprint her
name.
Recall the name he must --
the fathomed depths
resounded with the eloquence he craved
essence of truth, the knowledge
unashamed
this natal sequence that each man
acclaimed.
His love was LIFE...
like poetry her breath
that countermanded litanies of death
a flowered face that beggared
Eden's snow
its radiance timed -- a glimpse
of immortal glow.

ANOTHER ISLAND, CALLED MANHATTAN

The Sicilian hills melded into mosaics
  of gray and gold,
  as the ship pulled away,
Rich, with the lore of Grecian temples
  reciting the beauty
  of the Acropolis
  the sigh of Parthenon perfection --
Olive trees, twisted scraps of languor,
  deep in drought-despair, remnants
  of promises of other years,
Fruits, once luscious reminders
  of Eden's bounty, now caricatures --
  bent, like old men,
  their shrunken stems.

The poster had shouted, "AMERICA!
  Green, green!
  Sail to the Land of the Free.
  A new beginning!"
A sob sliced his throat --
  his sixteen years
  heavy as an anchor to his thoughts,
  proof of the manhood within him
  uncertain of the future,
  but defying the past.
Paisano! The rough cry startled him
  back to reality, a command
  which carried him
  to the top of the rigging,
  checking ropes,
  seagulls circling his giddy head.

(cont.)

He grew tall, a man-mast of hard work
responding to clarion calls
of the sails.
Nights were welcome reprieves.
He slept on a battered life-raft,
overlooking huddled slumberers,
escaped from the hold.

One starlit night, a child was born
on deck, the baby, squalling like
a hawser's rope straining
against the timbers of the pier.
"Water," cried the young mother,
from parched lips.
He ran to fill his cup,
held it to her mouth,
Smiled, whispering,
"A beautiful bambino,
and his name?"
the mother answered, "And yours?"

The morning after, sunlight streamed
through clouds,
then the tremulous cry resounded,
"THE LADY WITH THE LAMP."
America ...
He looked with proud possession of this
proof of welcome.
His new god-son was named "Mario."
A Sicilian gift for the Lady!

A CENTURIED BOW to EMILY DICKINSON
(1886-1986)

Highlighted beauty
    legended page
    her pen from inception
    half-child and half-sage,
Ambiance of tomorrows
    with outerworld hue
    the ineffable difference
    between pearl and the dew,
Flights of fancy while phrasing
    a provocative line
    that astral persuasion
    of thought rich as wine,
Starred syllables preening
    their space-world secure
    with moondusted meaning
    of skylark allure,
A century beckons
    this poet, whose scheme
    was to fashion with wonder
    a part of life's dream.

Helen T. Brown is a former chairman of Poets and Patrons and of Poets Club of Chicago. An accomplished poet, she is the national Chairman for Special Events for Pen Women (NLAPW).

Lola Morgan

POETS' ROUND TABLE

What brings us to this table?
Love of words, love of self,
need of others who are sentence-bound?
Word-petals drift through the room,
anguish explodes into meaning.
We listen carefully
waiting for our own poem to be read.
Some may be swollen with iambic
awareness
and some with unpunctuated freedom.
We almost smile about them,
but never quite.
Outside the room other lives
possess the selves which we pretend
we are.
We have hidden the notebooks
of loneliness
from our parents, husbands, children.
The old and the young are here,
strangers closer than lovers.
We come to this table hungry
and the bread of being together
will nourish our poems for tomorrow.

...Second prize, free verse form
in national contest sponsored by
Alexandria Branch of NLAPW in 1976.

OUR TINSEL SWORDS

We challenge time with tinsel swords
        and thrust
With vague pretense of valor
        at the years
Which have absorbed the brightness
        in their dust,
When passion dies and laughter
        disappears.
The petals drop from every dream
        and rose.
We gather them with old, arthritic hands
And find our fingers cannot keep
        nor close
On beauty past, on flowers
        from younger lands.
We know the sentences to welcome night
And words to greet a sunset or a star,
But we would hold the morning's
        early light
Which dimmed so soon, which now
        is lost and far.
Where are the joys and where was wonder
        spent?
Our tinsel swords are old and soiled
        and bent.

...First prize Shakespearean sonnet from Alexandria Branch, N.L.A.P.W. 1979.

ALL MY APRILS

The words you said had little, laughing
 wings,
They fluttered in my heart and stroked
 my mind
With lilac-fingers, joy that would
 unwind
Across the years, through other loves
 and springs.
To every April that lost April clings
With touch and taste which I can never
 find
Again and old wounds ache although I
 bind
The flesh with bandages experience
 brings.
We knew that ours was but a time
 between,
A blossom broken from a fruitless bough,
A song's beginning and its ending too.
But that old melody is sharp and green
As yesterday invades this careful now,
For all my Aprils still belong to you.

...POET LORE

THE PAINTED HORSE

It is a painted horse on which I ride,
The silver, gold and pink are chipped
    and thin.
Where are the meadows, green
    and summer-wide?
From what strange stable did this hunt
    begin?
I look for shining rings. They must be
    here
As once they were when love was new
    and gay,
The cotton-candy man must soon appear,
Perhaps he will recall my yesterday.
I clutch a ring of words but it is old,
As old as dreams which do not last
    the night.
This carousel is rusty, damp with mould,
A life ago I knew its brief delight.
The prize I snatch is dull, no sentence
    sings,
The horse goes up and down but has
    no wings.

...THE LYRIC

Lola Morgan was born in Dayton,Ohio, and has been a prolific poet. She moved to Chicago and joined the Poets Club in 1954. She is now living in Cape Coral, Florida.

Sara Beth Marshall

VEHICLES

People in line for food
gathering clothes for the cold
a child plays
an old man praying
a youth studies
a woman lights a candle
baby gurgles as baptism water
wets its head
individuals share fellowship
pilgrimages for Christianity
groups gather for fellowship
corporate ritual of communion
children work with toys
a cheek slapped symbolizes purity
his peaceful gaze of patience
flowers in a cornfield
a doctor heals within faith
a child collects the offering
an eternal flame
the Alpha and Omega.

EVERY YEAR

Spring brings its own testament of life
through nature.
Birth of all beings, and plants' growth.
Individuals experience faith by
spiritual witness
of the Lord's gift to each of us.
Easter:
and its hope of Eternal Life.

SMALL OCCASIONS

A fresh flower for
your shut-in neighbors
and time
to hear of their
memorable holidays.

Seeing the children
on the block
leaping and hopping
through the biggest yard
hunting for their share
of colored eggs.

Smelling the fresh air
of spring moving the newly
budded trees and bushes
so lightly, as the birds
gather for fresh seeds
in the feeder.

Touching the fabric
of your daughter's special
new dress she made for
the celebration.

Tasting the treats
on the family table
for the holiday feast.

It is these small occasions
that can witness
the large events of life.

A RELATIONSHIP

A month ago,
    went away.

Yesterday:
    A china bowl
    slipped and smashed
    into jagged shapes
    and white dust.

Today:
    Painting finished.
    The easel is bare.

Tonight:
    Heart crumbled
    but,
    not away, feelings now surfaced.
    Again, alive.

Friendship still
    the
    center
    of reality.

Tomorrow:
    Alone,
    what does survival
    entail?

A month from now:
    Hope
    of what?

THE DANCE

there is a pole topped
topped with a wreath of flowers
ribbons attached in order
with pastel colors
young ladies position around
carrying their pretty streamers
as rhythmically they move
creating a Royal Diadem

if only all over the world in May
the first celebration
could be held this way.

Sara Beth was raised and educated in Pennsylvania, is a graduate of Charles Morris Price School of Advertising and Journalism. and has lived in the mid-west 27 years. She served on the Board of Illinois Women's Press Association, is a member of National Federation of Press Women, and the National League of American Pen Women, Chicago Branch. She has done free-lance children's text book materials.

Rita Duskin

HAIKU

Life is a flower
Fragile and fragrant, fading
As we breathe on it

Each day I awake
And each day begin again
How long the road!

Like a good mother
The hot sun blankets me
I know what love is

Watching spring unfold
Within me I feel sprouting
Tiny seedlets of song

There is still today
And tomorrow fresh with dreams:
Life never grows old

AT NIAGARA

What am I
Against these thundering Falls,
But a transient atom
Seeking certainties;
A frail and foolish voice
That vainly calls,
Asking an answer from the centuries.

BE NOT SO PROUD

Be not so proud
my lovely:

You too
will grow old
Your skin fold
Your heart cold

Your tale told.

OLD AGE

We bumble
and fumble
and mumble
and grumble
and stumble
and tumble

Then -- finally

cr

um

ble

THE MOVING FINGER

Experience writes its memoirs on my face
Impossible to erase

FRESH FRUIT

When I taste raspberries,
again I am a little girl
climbing the 102 steps
at Spring Lake, Michigan,

Playing Blind Man's Buff at twilight,
stealing berries at dawn
better than Hansel and Gretel's
or any I have tasted since.

Again I catch my glimpse
of magical hills and rainbow waters
and unbelievable spaces and places
beyond the concrete world I knew.

IT'S SPRING

There's a wanton rhythm in the trees,
Sap is swelling in the veins
Let out the reins!

REMINISCENCE

Hymn to the Sun!
Clarion music
calling for naked dancing
In the morning of life

DEDICATION (To Boris, my husband)

The years have knit us closely;
Stitch by stitch, the garment made
Of pain and beauty, romance, work,
        and love.
To rip us now apart would be
Like night without stars,
Day without sun.

Rita Duskin was a former Chicagoan and member of the Poets Club of Chicago and Poets and Patrons. She died in February, 1986, at her home in Hallandale, Florida. Nominated for Florida's poet laureate, she was the author of THE FRUGAL CHARIOT (Windfall Press). Her articles and poems appeared in the INTERNATIONAL ROTARIAN, SCIENCE OF MIND, and other periodicals, newspapers and anthologies. She was a member of the Academy of American Poets, National Federation of State Poetry Societies, and Florida Poetry Society.
A poetry educator as well as an award-winning poet, Rita Duskin spoke to clubs, libraries, senior centers and schools in Chicago, Florida, and Phoenix, Arizona; compiled and edited a children's poetry anthology; and presided over literary groups. All of her poems printed here except the haiku are from THE FRUGAL CHARIOT.

Esther Kossoff

BEFORE ENTERING A HOME FOR THE AGED

My clothes are all packed; I'm ready to go.
Live with my children? That isn't my way!
If they had *asked* me, I would have said "No."

The woman who helps me - - she's well, "so-so";
She does her job, but has little to say.
My clothes are all packed; I'm ready to go.

I don't...don't know why my spirits are low;
They don't need me now, as once, yesterday.
If they had asked me, I would have said "No."

When they were little, from head to their toe
I cared for them, nourished them -- and now they...
My clothes are all packed; I'm ready to go.

They really are good. They took me to show
A clean, pleasant place to live and to pray.
If they had asked me, I would have said "No."

I saw a friend there; it won't be too slow...
The house of children is nowhere to stay.
My clothes are all packed; I'm ready to go.
If they had asked me, I would have said "No."

TO TAKE THE TIME

Oh, let me take the time to stop
and stare,
To see the world as if through
newborn eyes,
To greet the early rose
with glad surprise,
Intoxicated by the gift I share!
Waves tousle weeds, and breezes
comb my hair.
Though all is brief, I know,
and each one dies,
Today I live! And so I rhapsodize!
"I loaf -- invite my soul" --
Walt Whitman's fare;
And though I work, and work is good,
I've found,
I pause to view the first brave crocus
bloom,
And laugh as sparrows chitter
in the trees.
I revel in a sight, a touch, a sound:
I hear a friendly voice that warms
a room --
I take the time for life's
small ecstasies!

...SOUNDINGS

EYE OF THE BEHOLDER

Everyone said she was a good woman.
Whatever the emergency:
a sudden illness, a new child,
she could always be counted on to help.
--But she was so cool, so aloof!
If I only touched her hand,
she seemed to recoil!
She couldn't fool me! I understood her!
She was wrapping herself up
in a blanket of "Good deeds",
trying to feel some divine warmth,
rather than that of mere human beings!
I could imagine her praying every night,
asking God to reserve a seat
in Heaven, close to His side.
I imagined her adding up each fine act
on her spiritual computer,
making points!

And today, -- at her funeral --
many have come.
They will be speaking words of praise,
but there will be very little sign of grief,
no deep sadness,
Surely they are here out of
a sense of duty,
the same as I am.
With a knowing smile, I remark
to a neighbor standing near me,
"I hope she finds her place
in Heaven. She certainly
worked hard enough for it!"

Why is she staring at me so strangely?
Why are all these people crying?

SPRINGTIME IN THE PARK
AT THE WATER TOWER, CHICAGO

In Spring, in the Spring,
  bag ladies come out.
One rests wearily,
  gazing about.

The bench where she sits
  is facing the street,
and three shabby bags
  lean next to her feet.

The bag lady looks
  with vague, veiled eye,
watching those other
  bag ladies go by --

shopping bags swaying,
  as they saunter through
the elegant stores
  on the avenue.

Esther Kossoff is a former news editor of the Lincolnite, columnist for the CHICAGO DAILY CALUMET, and English teacher. Among her credits are ORPHIC LUTE, BARDIC ECHOES, NIT AND WIT, LERNER newspapers, GARLAND COURT REVIEW, SUNSHINE MAGAZINE, CHICAGO CATHOLIC and EVANSTON REVIEW, and in anthologies AVENUE OF GOLD and SOUNDINGS. She graduated from U. of Illinois, where she was on the tennis team.

John Froehlich

SELF-PORTRAIT

A mirror sits for him.
He need not once instruct her.
She moves without a word
to satisfy his slightest wish.
Eyes stray with nervous frequency
from face to canvas.
At last, the masterpiece is done.
His painting hand
has told some small white lies.
His friends, observing,
easily understand
how he and truth
could compromise.

JOHN

It is a name I like.
It's solid, brief and masculine.
For 1900 years and more,
how many millions
bore my name with pride!
And then it was demoted
to a toilet,
and finally
to a hooker's customer.
At least they had the decency
to take away
the capital J.

THE HERMIT CRAB

He's made
a hollow mollusk shell
his home.
As he outgrows this shell,
his stalk eyes
search the shore
for something bigger.
Before he finds
the perfect fit,
his search
is turned to tedium.
This creature
thinks he is a large,
but he is just
a medium.

IN A NEIGHBOR'S BACK YARD

There is a bird bath here,
with Mary, mantled in her blue,
standing on a pedestal
centered just above the surface.
A dozen sparrows
flock around her feet.
Their bathing and their drinking
are partially pretense,
as instinct pays a homage
that's too deep
for human sense.

PAINTING BY AN 18th CENTURY MASTER

The white-wigged nobleman,
with exquisite tenderness,
barely touches with his lips
the maiden's hand,
extended so demurely.
Blushing,
she lowers modest lids
and bends her head.
He hopes it's an approving nod.
Cleavage between voluptuous breasts,
half-hidden
by her low-cut hoop skirt gown,
challenges his resolution
not to stare.
His single kiss
has lasted for 200 years.
He must believe persistent gallantry
will, in the end,
secure her heart.

John Froehlich has written seven books of poetry. His last one, by Lake Shore Publishing, consists of 93 illustrated limericks. Between writing and bridge, his second love, his retirement is never boring.

Louise Hajek

NO DEFENSE

She knew that look. The far-off places
tugged
At him again. Sometimes she wished that
he
Would drink or even beat her. Still she
hugged
The thought of sober kindness close.
Maybe
She would not care too much if he began
To loaf in town and eye the girls.
Indeed,
A woman worth her salt could coax a man
Back quick enough, if she but knew his
need.
The firelight limned his face; she saw
him stir
And felt a small, sad pride knowing his
eyes
Would never lift to anyone but her;
Then wept because she never could be
wise
With woman's word or way to ease his
load.
What weapon could she use against
a road?

...PORTLAND OREGONIAN, 1964.

GOOD WINE

The room is quiet, soaring hearthfire
        sings
And makes an etching of your wintry
        hair.
Watching, I tally all the precious
        things
That changing only change to grow more
        fair;
The dimple that has deepened to a groove
Of tenderness, the laughter marks now
        cleft
To kindly lines. Gently you sigh
        and move
To touch me and I note the years
        have left
Dark mottlings on your strong beloved
        hands.
You called them "cemetery flowers"
        and smiled
(Eyes bluer still in time's deep
        shadow-bands)
With all the fearless candor of a child.
The fire dies down, but I am warm
        and filled
With rich contentment like
        a rare tokay --
Good wine, the wine your presence
        has distilled,
Its vintage finer than the wine of May.

WAITING CAT

In lethal calm
the sleek arpeggios
of muscle hold.
With metronomic tail in play
the eyes of gold
are point
and counterpoint
of prey,
hypnosis bound --
Assured one error
will launch this instrument
of terror.

...THE LYRIC, 1954
...NEW YORK HERALD TRIBUNE, 1954
...BLOONSBERG MORNING PRESS, 1964

Louise Hajek began writing poetry over 40 years ago. During her writing career she won many awards for her poetry in local, national and international contests. Her work has been published in literary magazines, popular magazines and in school texts. She brought out three collections of her poems: ONE CANDLE, FROM THE TWISTING BRANCH, and REMEMBERED MUSIC. Her work includes everything from haiku to Shakespearean sonnets. The content ranges from religious themes to verse for children to humorous limericks. She died on May 28, 1985.

Don Cornwell

UNGROOMED

She rang the stable to announce
    She'd want her horse at ten.
The honeymoon had been great fun,
    But Oh! to ride again.
Arriving booted, dressed and capped,
    She sensed a change in store,
For things were clearly different there
    From how they were before.
Her horse had not been readied yet,
    Not brushed, not tacked, not tied;
And when she found him in his stall
    A note was pinned outside.
It said that all the help was gone,
    And she could now assume;
Although just wed she'd henceforth be,
    A bride without a groom.

...*THE CHRONICLE OF THE HORSE*

NEXT TIME

Come along, young man.
Pick up your bat and glove.
Wipe those tears.
I know the Bulldogs won, and that you,
only you, might have saved the game.
I'd like to say it doesn't matter,
        but it does.
I'd like to say there'll be
        a next time,
But there won't, not just like this.
But there will be other times
when a single hit will win the game.

(cont.)

You will get that hit, believe me,
next time.
I'd like to say I understand;
perhaps I don't.
Now come along and wipe those tears,
And I'll wipe mine.

THE THIRD DAY

And so the earth brought forth
in green array
Fresh growth of plants and trees
with fruit entwined,
Each one with seed according
to its kind,
As night and morning came
to the third day.
To man belonged the trees of fruit
that they
Be food except, forbidden and enshrined,
One called the Tree of Knowledge
of the mind,
Whose fruit would take all innocence
away.
When now in pain and fear our lives
begin,
And then by sweat of brow we earn
our bread;
Who has not wished that life could be
as sweet
As it was then before man tasted sin.
Our hope may be to know in years ahead,
What fruit it is of which
we should not eat.

NIGHT OF THE FLIGHT

At ten the flame of lightning came,
And the sky turned into day,
And hail beat on the window panes,
And cast its icy spray.
The huge plane rocked and then it fell,
And then it climbed again;
And then it fell, and then it climbed,
And then began to spin.
The lights went out and engine noise
Was hushed by rushing air.
The whistling sound warned
of ground, or sea, if sea be there.
And then in space where nothing is,
There was an eerie ring.
A starboard engine flashed with fire
And tumbled from the wing.
The screams of desperate fear began
In bass and high pitched tones,
And all the screams were harmonies
Of piercing shrills and groans.
Through clouds we slipped, and then
We dipped, and pointed for the sea.
We bowed our heads, and grabbed
Our knees, and knew eternity...
We struck the sea at sonic speed,
Like a sharpened saber tool;
And held the angle divers keep
Who do not splash the pool.
There was a splash, but was no crash,
No watery demise;
And then the nose was pointed up,
And we began to rise.

(cont.)

From depths of who knows just how far,
The plane swam for the air;
While not a sound was heard inside,
Except the words of prayer.
We surfaced like a rocket shot,
Then fell back to the sea;
And gently then began to float,
Away from tragedy.

CANINE GRAMMARIAN

Our dog knows English, one might say,
He's learned to lie down, not to lay.
He's found he's different from our cat,
Not different than, he'd not find that.
Sometimes he's wished he were like us,
Not wished he was, that's ludicrous.
He cringes when he hears "he don't."

He doesn't like it and he won't.
He's heard us talk of cat and he,
But knows that "him" fits properly.
He barks to say, "May I go out?"
There's never "Can I" in his shout.
He and myself, "myself?" Oh my!
He'd frown because the word is "I."
Some say he's not "much of a pup."
He'd like to tell them, "Look that up."

All of Don Cornwell's poems included here are from SENSE AND NONSENSE. A poet for more than twenty years, he is past president of Poets Club of Chicago and Poets and Patrons. His light verse has been regularly published in the CHRONICLE OF THE HORSE, a national horse magazine, and he has completed a storybook poem, HORACE THE PONY AND YOU.

June Shipley

SEASONS

When seasons change I hear
The wild seductive call
That bids me follow; lures me back
To that forgotten place where life began,
And I, shaped from a rib of the first man,
Left all, and listening to that call
looked for adventure on primeval paths.

When seasons change and migratory instinct
Pulls me, entices me, tells me to seek
The Eden I once lost,
I must stand firm, remembering the cost
Of restlessness, and how --
When seasons changed back there;
And I knew where
The apple tree hung red with fruit,
A whisper from the grass said, "Do it!"
And I listened to it!

When seasons change and the wild winds
And errant clouds disturb me,
I close my eyes so that I cannot see,
And ears, and do not hear
When they call out --
For I, remembering "when" am not about
To go through <u>that</u> again.

...PROFILES CUT FROM THE WAVES, 1983

WHY NOT?

I do not care why Caesar bled,
Or why Sir Walter lost his head,
I do not care why Custer died,
Or why the Inquisition fried
The unbelievers -- nor do I
Give a fat fig for reasons why
The shift of power is so bloody,
I do not study botany, economy,
philosophy.
Or care why Roman legions built
Great roads and aqueducts in regions
Far from home and as for Rome --
I do not care if it was sacked.
In fact, I do not care if Greeks
Fought wars -- What's more I do not seek
To know the reason why
Narcissus turned on Echo.
I could care less about Prometheus --
Who thinks about the riddle of
the Sphinx?
Not I -- I know the past -- is past.

I wonder if my cash will last
Beyond my time and I'm concerned
About my income tax return,
I worry if my car won't start,
Or if my washer falls apart,
I care about my aching tooth,
And wonder what's become of youth,
I care about the company
I keep and when I cannot sleep,

(cont.)

I care about that -- dreadfully.
I care what politicians say,
I do not look the other way
When they start fishin' in my purse,
Or worse.
I care if it is cold or hot,
For relatives -- and quite a lot
For cats and dogs and flowers and bees,
And beaches, lakes, and mountains,trees,
I cannot bear the wrongs I see,
Fanaticism, bigotry,
Old people have my sympathy,
But most of all --
I care for me.

ODDS

Bloodlust astride,
And the awful sound
Of the hunter's horn,
And a pack of hounds,
And a hundred hooves,
Pounding on the ground
And
    one
        small
            fox.

RESIDENT, OCCUPANT, SPOUSE

Where the vacuum cleaner goes,
There go I.
Where the shy dust-bunnies grow,
There go I.
Where the crumpled bed-clothes cry,
Where the feather dusters fly,
Where the dirty dishes lie,
There go I.

Where the laundry starts to call,
There go I.
Climbing up the kitchen wall,
There go I.
Guess I'll just ignore them all --
Where's the nearest shopping mall?
Hi-dee-ho and h-dee hi!
There go I.

...First Place Award, Poets & Patrons, 1985

June Shipley is active in several poetry groups in Illinois and Indiana. She also writes short stories. She is president of the Lake County, Indiana Poetry Club and treasurer of the Indiana State Federation of Poetry Clubs. She has appeared in several anthologies, newspapers and Purdue University's FINE ARTS magazine.

Blanche René

CHICAGO BIRD SAID

My parents built a nest for me, a clay
cradle of twigs and string
                    left from a kite.
At incubation time, I couldn't stay
a yolk of yellow in a sea of white.

I've flown above the Michigan Avenue
                         bridge
both when it's down and raised
                    for boats to pass.
I've flown above the city of Park Ridge,
a short flight from Chicago's
                      Loop of glass,

and over the tallest building
                         in the world
a few roof gardens from
                 the State Street Mall...
Sometimes, I land on grass the night
                         has pearled,
sometimes, upon a distant village hall.

I'm fortunate to have a wishing bone,
but I can't fly when ice forms
                         on my wings.
I'll migrate southward to a warmer zone
far from the howling winds
                    that winter brings.

...ORPHIC LUTE

VOICE OF THE TURTLE

I came to life hatched
by the sun's warm rays,
stepped from the nest my mother
dug in the sand.
Now I'm excited about my turtle days
of living competitively
in creatureland.

I've built a house designed for only me.
With precious load, the lifetime house
I haul,
I walk the land. I paddle in the sea.
Faced with foe, I make
a quick withdrawal.

Don't move, play dead! On this defense
I'm sold.
In nature's wondrous ways,
I put my trust,
enjoying life, the art of growing old
until death turns my mobile home
to dust.

...EARTHWISE LITERARY CALENDAR

Blanche René has had prizewinners in regional and national contests. She has given lectures on poetry, has judged for poetry book awards and for poetry contests.

THE HAWK AND JANE

For contrast, let's compare Jane
to a hawk,
But only in a locomotive way.
Each has two feet to hop about and walk,
So, in that sense, they're equal,
I would say.

The hawk's no dunce, but can it ride
a bike,
Roller skate, or skip-the-rope
like Jane?
It's fair to mention, she can't
solo like
The hawk who flies above both
hill and plain.

And then again, the hawk can't fly
as far
Or half as fast as Jane,
for don't forget
She flies around the earth,
a planet-star.
Jane flies in comfort in a jumbo jet.

Blanche René has poems published in SOUNDINGS, an anthology of poetry; POETS OF THE VINEYARD; and in ILLINOIS AUTHORS COOK BOOK, (both culinary and literary).

THE INNER ROAD

We travel to the moon, our satellite.
We fly to all the corners of the earth.
Sometimes such travel helps us
in our flight
From mundane cares to thought
of higher worth.

The Inner Road we take from sense
to Soul --
That's the journey that makes
the spirit whole.

...FOOD FOR THOUGHT,
... BUDDHIST JOURNAL (Bangkok.)
... ILLINOIS AUTHORS COOKBOOK.

Blanche René was a member of the Society of Midland Authors, O.C.W.W., Children's Reading Round Table, National League of American Pen Women, and Poets and Patrons, Inc. She was secretary of the Poets Club of Chicago, and was currently president of Poets & Patrons Inc. She was also a liaison from Illinois to the Board of Directors of EARTHWISE PRESS, Miami. Blanche is author of two books, A PONY CART OF VERSE and BOOK OF THE REBORN, London. Blanche died in 1986.

Evelyn Schmidt

MOVE

Move in this miracle of chemicals,
this independent breath,
with discs of memory to play
and a needling word or fragrance
to start it playing;
not even needling need it be,
a faint fragrance,
an almost imperceptible sound.
Escape now,
move out, move up, move away.
Grass is greener in a far field.
Food tastes sweeter at a foreign table.
New voices in the ear awake a sleeping,
a walking stirs the mind,
and laughing eases tension.
Move, yes, but to a warm
field of memory;
a remembered feast,
a song, a whispering "I love you";
a walk under the sun through
a golden wheat field
where swaying whiskered heads
prick and caress the thighs
and red-winged blackbirds flash upward
like a sudden cloud casting shadows
where bell laughter floats on the wind
and looking far, the horizon shimmers
in the sun, yes, move.

COMPANIONS OF MORNING

Over the Nile rises Ra
with bright awakening rays.
Tutankhamen thanking Ra for the light
meditates; flood receded, crops planted,
Ra brings forth abundant food.
Tutankhamen walks
the beautiful river shore,
enclosed for the royal family,
never blackened with rich soil
carried by the powerful Nile.
He seeks Ra's companion of the morning,
the tall heron.
There, taller than the rushes
he stands
on one slender leg.
He peers regally in command,
a great monarch,
and like the sun,
he sets the temper of the day.

MOONLIGHT NIGHT

I feel the dew between my toes,
and lying on the grass, I watch
a small white cloud.
Shadows come slowly, creep closer
through the full-leafed maples
and the moon doubles the silver.
Daisies and roses soften, buoyed up,
floating in the moonlight.
The farmhouse blurs
while the two patios, lit by
the night, and the chimneys
move plainly, securely
toward the sky.
I hear the walk of horses,
and their blowing on the water tank.
I see their shining flanks; a white
horse ghostlike in the moonlight
leads them to the pasture.
Crickets create their sound
of sameness,
drowning out other sounds,
and I drift off to sleep.

BLUE OCTOBER

When morning shines above the mist,
A sky that blue October knows
Is shouting to a boy that goes
Into the forest, "You exist
And with the squirrels you can tryst
Among the leaves." The young boy glows
When morning shines above the mist
In sky that blue October knows.
The geese fly southward. They insist
The day is made for flying,
                    thinking snows
Will come and where clear water flows
There will be ice. A boy cannot resist
When morning shines above the mist
In sky that blue October knows.

Evelyn M. Schmidt, writer, artist, and retired teacher, has been editor of the Brookfield Enterprise Poet's Column, The Inland Journalist, and Chicagoland's Award-Winning Poetry 1972 and 1973. Among her most treasured prizes, she counts "First" in the Poetry Workshop of the Mississippi Valley Writers' Conference and "First Prize" in the Free Verse Category of the Chicagoland Poets and Patrons Contests. Much published in traditional verse, Mrs. Schmidt is also an haikuist and appears regularly in GEPPO, the International Haiku Newsletter. She is a former president of Poets and Patrons.

Orrin Singleterry

ARTIST AND AMATEUR

Carlos Montoya, I have heard you play
Your wild flamenco music on guitar.
Chords crash and echo, rhythms
        strut and change
In Spanish gypsy frenzy, while deft
        fingers
Pluck out a melody like rushing streams
Or burst of birdsong.

        This is my guitar.
It has six strings like yours,
        and nineteen frets,
I practice scales and chords and
        stumble through
Some simple exercise. Often it seems
That I forget as quickly as I learn.

Why do I keep on trying, why presume
That mastery is possible, why dream
That this guitar may have a gypsy soul?
Carlos Montoya, I have heard you play.

WILD RIVER

Wild river, ever rushing boisterously
Through foaming rapids hemmed
by canyon walls,
You leap with bold abandon to be free
When cliffs impede your path
at waterfalls.

What pent-up energy, what wanton force
Unharnessed, wears these naked rocks
away!
White water, whirling, swirling
on your course
Trumpets wild water music
through the spray.

You come from raging storms
and melting snow
In the high ranges; through
the arid land
You carve a headlong canyon course
and go
Swift to the ocean on some lonely
strand.

You find your destiny, but not an end.
Such virile energy can never be
Entirely spent, but ceaselessly
will send
New waves and currents through
the restless sea.

THE FOUR SEASONS

When lilting spring turns grass from brown to green
While robin, wren and catbird sing-along,
And mist of early willow bud is seen,
My mutant voice may be beguiled to song.

When lazy days of summer then appear,
While bud to leaf and flower to fruit progress,
And mower and cicada dull the ear,
I take my ease in comfort, more or less.

When autumn strews the ground with leafy quilt
And children trudge reluctantly to school,
I contemplate my castle still unbuilt
And chide myself for being but a fool.

Now must I brace against the wind and cold,
For winter comes to tell me I am old.

FAIR AND WARMER

The weather man holds forth each night
With pictures from the satellite,

With isobars and highs and lows,
And who is hottest, and who froze.

The East has storms, or "thorms", or "sneet,"
The West is wet, the South has heat,

An arrow of Pacific air
Will bring a change to warm and fair.

The temperature is so-and-so,
The barometric pressure's low.

I finish off my bedtime snack
And go consult my almanac.

Orrin Singleterry was born in 1902, raised in northern Illinois, graduated from Aurora College, served as a missionary to China, and was editor of a religious weekly. He held the position of treasurer and business manager at Aurora College, and served as treasurer of Poets and Patrons. He now resides in northern Florida.

Lillian Foster

CHICAGO

Is there some magic that has made this spot
The rendezvous for teeming hordes of men
Who mill about, their speech a polyglot,
And make our town sometimes a robber's den
And sometimes too a Mecca for the best
Of citizens from all round the earth?
I think the magic lies in our conquest
Of many attributes, culture and mirth,
Our town has everything for everyone --
Here vice and virtue meet, sometime in strife,
Sometimes to wink and pass as if in fun,
And knowing each to other gives new life.
Chicago is a lusty town; her voice,
Refined a bit, can make us all rejoice.

MEMO TO MY LOVE

Do not grow old, my darling,
with the years
That fly so fast and far beyond
our sight;
The time for growing old,
the time for tears
Must not be yet, until we find
more light.
We have not done enough to pay our debt
To life, whose mobile face has smiled
our way.
We have not done enough -- my darling,
Let us help send out more light
in this, our day.

EPITAPH

C'est fini:
The hate, the ugliness, the sin --
All these dull death has garnered in.
Here lies the shell of one who knew
No joyful way, no friendly view.
Let no one weep that life has fled;
Mourn not, but let a prayer be said;
For she, whose stone we place today
Believed in God and liked to pray,
Yet felt no love for humankind;
God rest her soul, let Judgement find
Some good that mortals did not see --
C'est fini.

NIGHT FLIGHT

I flew into the sunset's rose and gold
And night was slow to fall. I followed light
Reached out to beauty there that I could hold,
As long as life should last. I gained new height,
Dropped off old fears that once had fettered me,
Resolved to find new purpose, though the dark
Would hide my destination. I must be
About some useful work, retain the spark
Between me and infinity that fear
Should never touch. Here God devised with grace
A sunset, filled with mystery, so near
That self was selfless, caught in wondrous space,
To glimpse a solar knowledge more profound
Than any known by mortals still earth bound.

AUTUMN

Today the world is glowing, as the fire
Of Autumn, lambent, sweeps
across its strings;
Its beauty makes my heart a golden lyre
Whereon is played an aria that sings.
The red-gold Autumn hints
of coming cold;
The thudding sounds of falling
fruits and nuts
Join rustlings of dried blades
of corn to hold
The rhythm of a dying year that cuts
Deep into hope. One fears that life
dies now,
That Winter marks an end, a final death;
But faith and other Springs
will not allow
Autumnal song a dirge-like
tone or breath.
I will reach out, embrace
this brilliant day,
And hold its beauty close till
budding May.

Lillian Foster is a former president of Poets Club of Chicago. Sad to note, she is suffering from Alzheimer's and is in another state. We wish she were able to see and enjoy this volume of poetry.

Agnes Wathall Tatera

THE LORELEI REFORMED

Don't set your will to cross the stream, my love,
when I stand opposite and waiting.
The thinning, thicking mists swirl from my eyes.
My lips are traitor to my words, and baiting.

Our hands seem close enough to touch, my love,
but waves ride treacherous in the narrows
and if you trick a path from rock to rock
you'll find them mossy stepping-stones to sorrows.

So kiss me only with your eyes, my love --
then turn your back on love's confusion.
I take no dark delight in drownings, love:
my song is powerless, and my spell illusion.

...THE LYRIC

VARIATIONS ON A WELL-KNOWN THEME

There's old Mrs. Komiskey,
probably slightly high on whiskey,
taking out the garbage and scattering some bread,
her lips a tantalizing red.

She's a nice lady -- let there be no confusion
and she gives no houseroom to illusion:
in her housedress and sweater she's no overripe queen.
She knows she's built like a washing machine

with a sloping facade, but she's got every flavor
of lipstick and gloss and lip frost to savor
that ad boys and chemists with their heads together
could dream up for girls with brains like a feather.

A color for all the fads and fashions
and sly hints of fantastic passions:
Cherry Frisson, Pink Titillation,
Tawny Trip and Silky Sensation,
Hickey Purple and Funky Flame.
The ad boys don't ask, "What's in a name?"
And Mrs. Komiskey's got them all,
even Natural Natural.

Truth is, she's using up her daughter's,
along with her perfumes and toilet waters.
Now, there's a girl that's seldom around,
so her Ma doesn't get a chance to ask it,
just picks out cosmetics from the waste basket.
She wipes them off with a piece of cloth,
sprays on the scents and paints on the gloss
to cover up her ache and loss
in a simple combination of
economy with an act of love.

...A TRICK OF LIGHT

DISARMED

Neville Chamberlain
is not the only one
who trusted an umbrella
against a rain of fire.

I, too, have sought
peace in our time
against your unpredictable moods,
carried a collapsible shield,
a small petunia-domed gazebo,
shelter enough for reasonable gusts
and showers. My umbrella --

Oh, but *your* rabid weathers;
your stinging, slanted downpours.
Before I can think to close the silly thing
-- use it as a weapon --
it blows itself inside out,
useless in your storms.

...ORBIS (Great Britain)
...A TRICK OF LIGHT

THE APOSTLES OF ELVIS

Not everyone has powers
or opportunity to raise the dead,
but these reincarnaters everywhere
offer their youth and verve
to prop the saint decayed.

Impulse to nourish,
impulse to be fed by the idolaters
prod them to trade their own identities
for another's.

The exchange is truly made:
faces various and numberless
are magnetized by the same passion;
the flexing voice and thighs
resonate as one.

So satellites,
reflecting dispersed energies,
might orbit
a burnt-out star.

...ARIEL IV, Triton College.

Agnes Wathall Tatera also writes under her pen name Agnes Wathall. She saw her first work in print when she was twelve years old. She publishes regularly in poetry journals and has been a prizewinner in numerous competitions. Her book of pocms, A TRICK OF LIGHT, was published in 1984. She is the current president of Poets Club of Chicago.

Kay Wissinger

ON LOOKING AT THE MONA LISA
(Second Thought)

The picture tantalizes. Is that smile
A teasing trick of nature she possessed,
Seemingly enigmatic, all the while
No riddle really there -- a curious jest?
Or is there hidden in those upcurved lips
And distant look about her deep grave eyes,
A secret we would share; as passing ships
Hail one another, far from native skies?
We are enamored by the mystery
Of this strange smile. Is there a hint of tears
Which give to it a haunting poignancy
Caught by the artist for the wondering years?

Or did he, with amused capricious guile,
Himself add mystery to an empty smile?

...THE CHRISTIAN SCIENCE MONITOR, 1984.

I KNOW, I KNOW

Little girl, you own this day,
Drowsy, warm, and tender blue;
Dreamlike it will ever stay
In your heart, enchanting you.

When sometime the shadows fall
In your life, as in the sky,
Suddenly you will recall
A white sailboat drifting by

When you stood upon a hill
Close beside me, looking down
Through the air, golden and still,
At the sea; the little town

Whitewashed in the summer sun
Nestling where the beach curves wide,
Fishing boats all set to run
With the turning of the tide.

As the golden moments drift,
On your face I see a glow;
You have just received a gift --
Little girl, I know, I know.

...THE LYRIC

THE LONE CYPRESS

Why does it turn its back upon the sea,
Is it the sighing that the wet winds bring,
A sense of loneliness and urgency
Caught in a seagull's cry and outspread wing?
Or is the call of earth too strong, and so
Bound to the sea upon this barren place,
It bends to where in hidden woodlands grow
Wild flowers that match the seafoam's spiral
grace?

Standing beside this rock which none may climb,
For touching it with hands of love might kill
This lonely tree defying ruthless time,
I feel the strength of its primeval will,
As rooted here above the wave-lashed sand
It leans toward the home-iness of land.

...1st prize award, Chicago Branch, NLAPW, 1980
...THE LOOKOUT.

A VIGNETTE

Wrinkled and spare, as light and dry
as desert air, she passed me by
in faded gown; seeming to me
of thistledown fragility.

Lonely her days go, one by one,
whose ardent gaze brought her a son;
An alien world about her now
presses its thorns upon her brow.

Vanished the dawn, and bright noon too;
Now life goes on in dusky hue
through hours that plod.
    I think of this --
One time a young god
    craved her kiss.

...*THE LYRIC*
...*First place in Illinois State Contest NLAPW, 1978*

Kay Wissinger was born in County Tipperary, Ireland, and came to the United Sates at the age of 13. She has been writing poetry all of her life and has been published regularly since 1940. She is an active member of the American Pen Women and Poets and Patrons.

Index of Poets

(Cont.)

Copies of this book may be obtained from any member of the Poets Club of Chicago, or ordered directly from the publisher.

Lake Shore Publishing
Attn: Carol Spelius, Editor
373 Ramsay Road
Deerfield, IL 60015

Price is $8.95
Postage and handling $1.00